This book belongs to

_____

Copyright © 2020 Emily J. Muggleton

All rights reserved. No part of this publication may be reproduced, distributed, or transmitted in any form or by any means, including photocopying, recording, or other electronic or mechanical methods, without the prior written permission of the publisher, except in the case of brief quotations embodied in reviews and certain other non-commercial uses permitted by copyright law.

First Printing, 2020
ISBN: 978-0-578-81789-7

Instagram: @emilymuggleton

SK1 Spacesuit

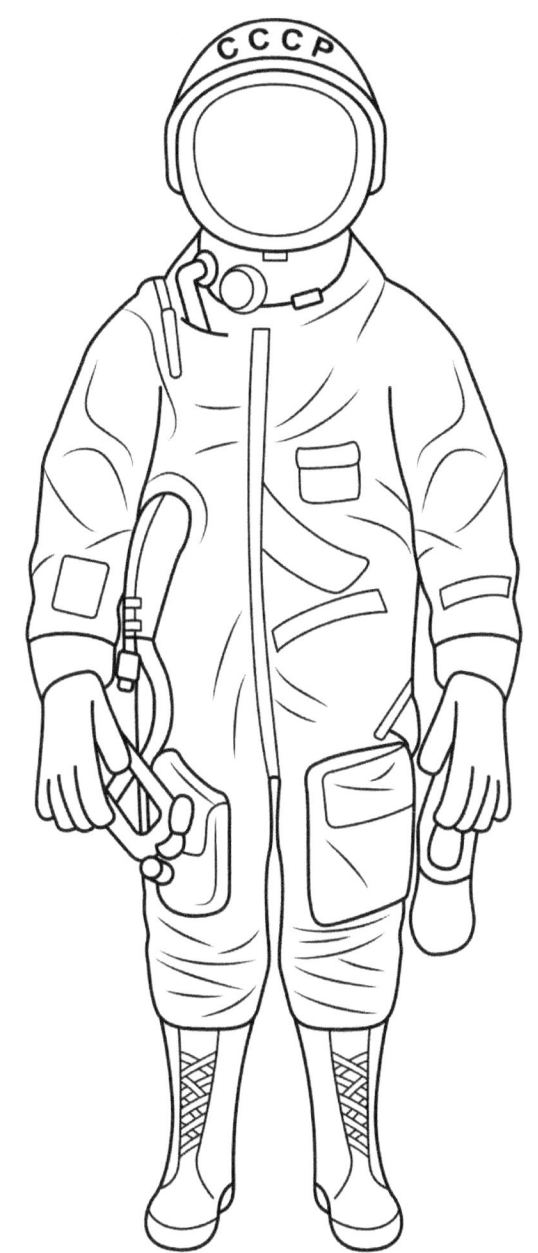

Origin: Russia
Date: 1961 - 1963
Missions: Vostok 1 - Vostok 6
Function: Intra-vehicular activity (IVA) and Ejection
Fact: First spacesuit ever used.

SK1 Spacesuit

Mercury Spacesuit

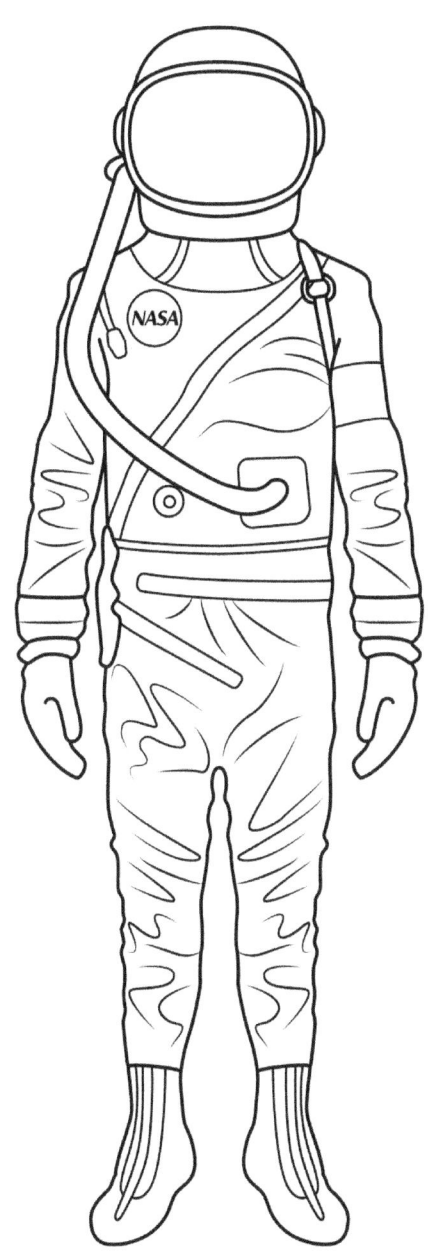

Origin: USA

Date: 1961 - 1963

Missions: MR-3 – MA-9

Function: Intra-vehicular activity (IVA)

Fact: Used for the first man-in-space program in the USA

Mercury Spacesuit

Berkut Spacesuit

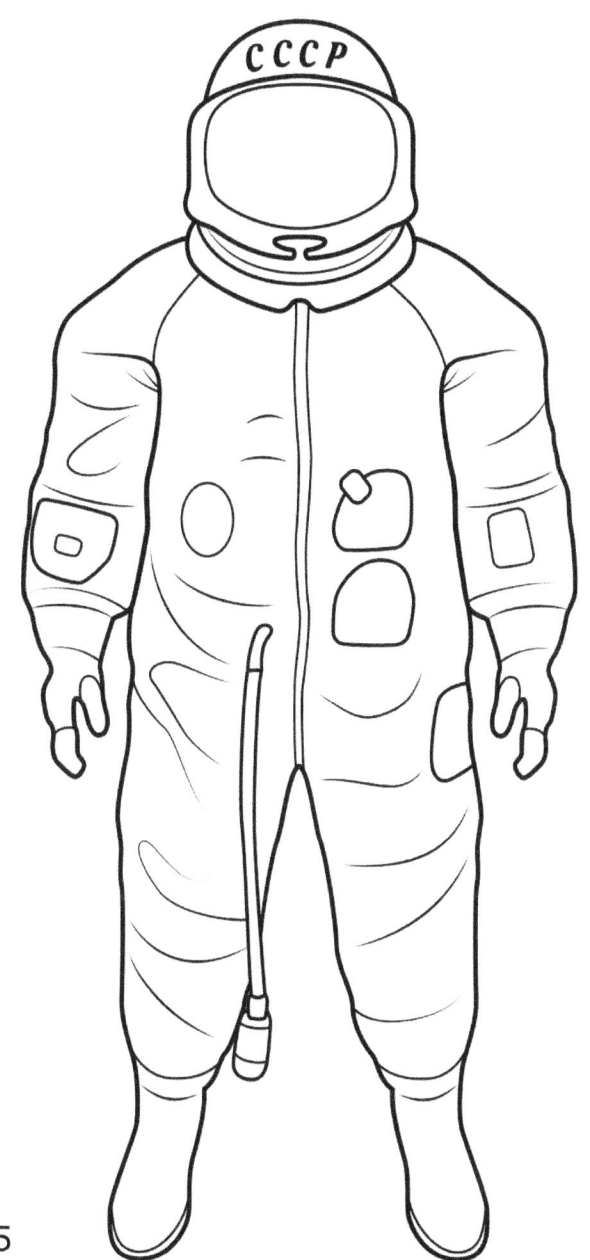

Origin: Russia

Date: 1963 - 1965

Missions: Voskhod 2

Function: Intra-vehicular activity (IVA) and orbital Extra-vehicular activity

Fact: Worn by Soviet cosmonaut Alexi Leonau for the first ever

Berkut Spacesuit

Gemini G3C Spacesuit

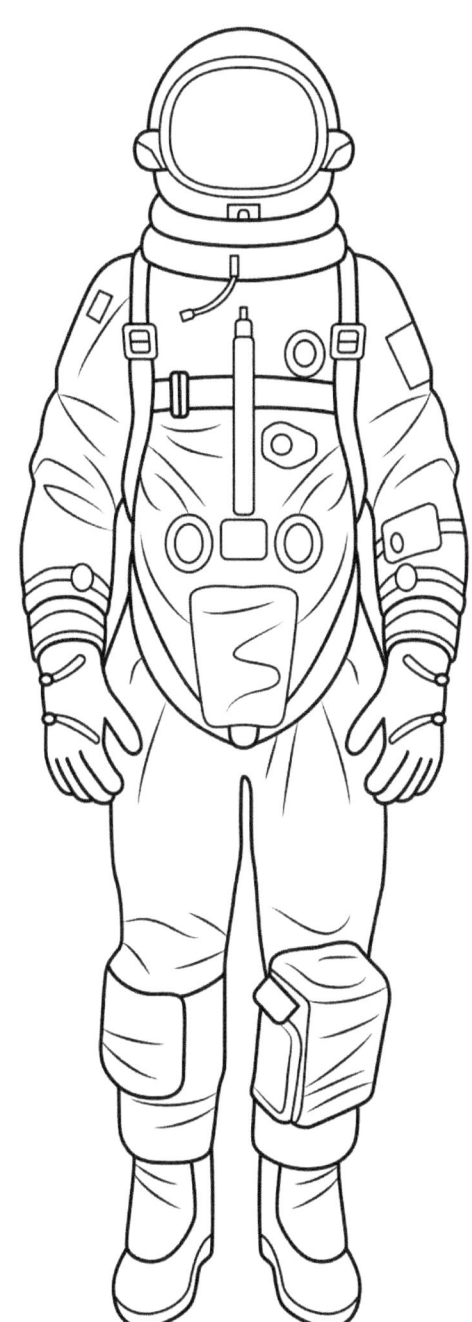

Origin: USA

Date: 1965 - 1966

Missions: Gemini 3,6 & 8

Function: Intra-vehicular activity (IVA)

Fact: Suit system included parachute and flotation systems to enhance crew survivability

Gemini G3C Spacesuit

Gemini G4C Spacesuit

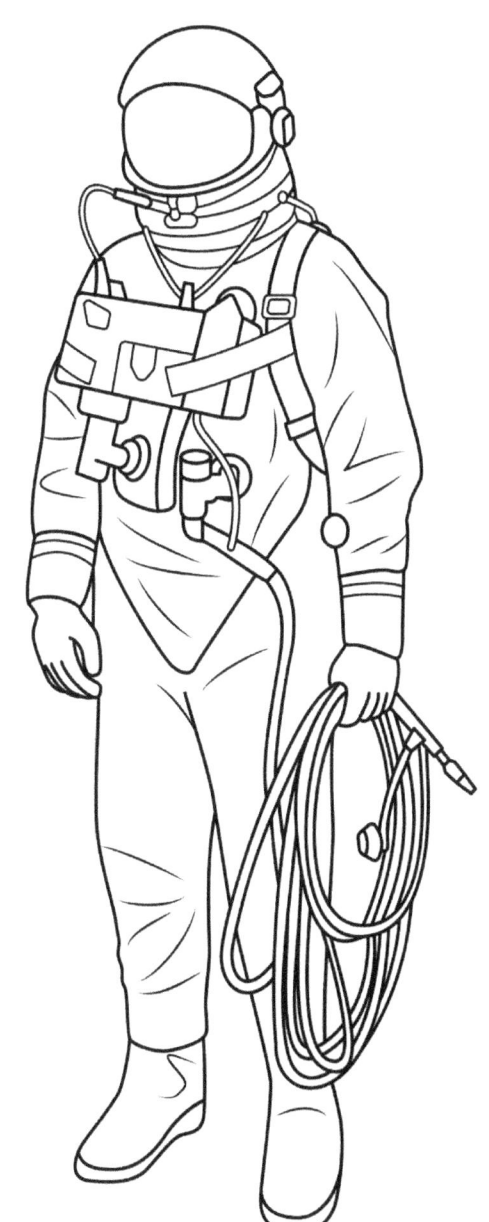

Origin: USA

Date: 1965 - 1966

Missions: Gemini 4-6, 8-12

Function: Intra-vehicular activity (IVA) and Extra-vehicular activity (EVA)

Fact: Worn by American astronaut Ed White for the first American space-walk

Gemini G4C Spacesuit

Krechet-94 Spacesuit

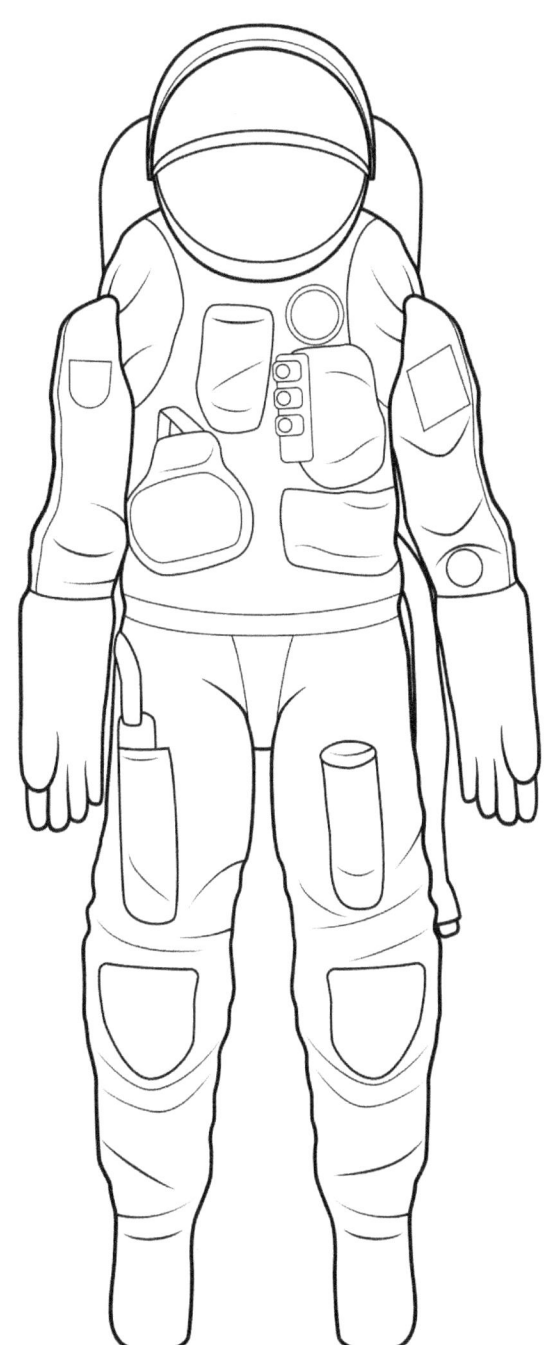

Origin: Russia

Date: 1967

Missions: Never Used

Function: Lunar Extra-vehicular activity (EVA)

Fact: Developed for lunar excursion during the Soviet manned lunar program

Krechet-94 Spacesuit

Apollo 11 A7L EMU Spacesuit

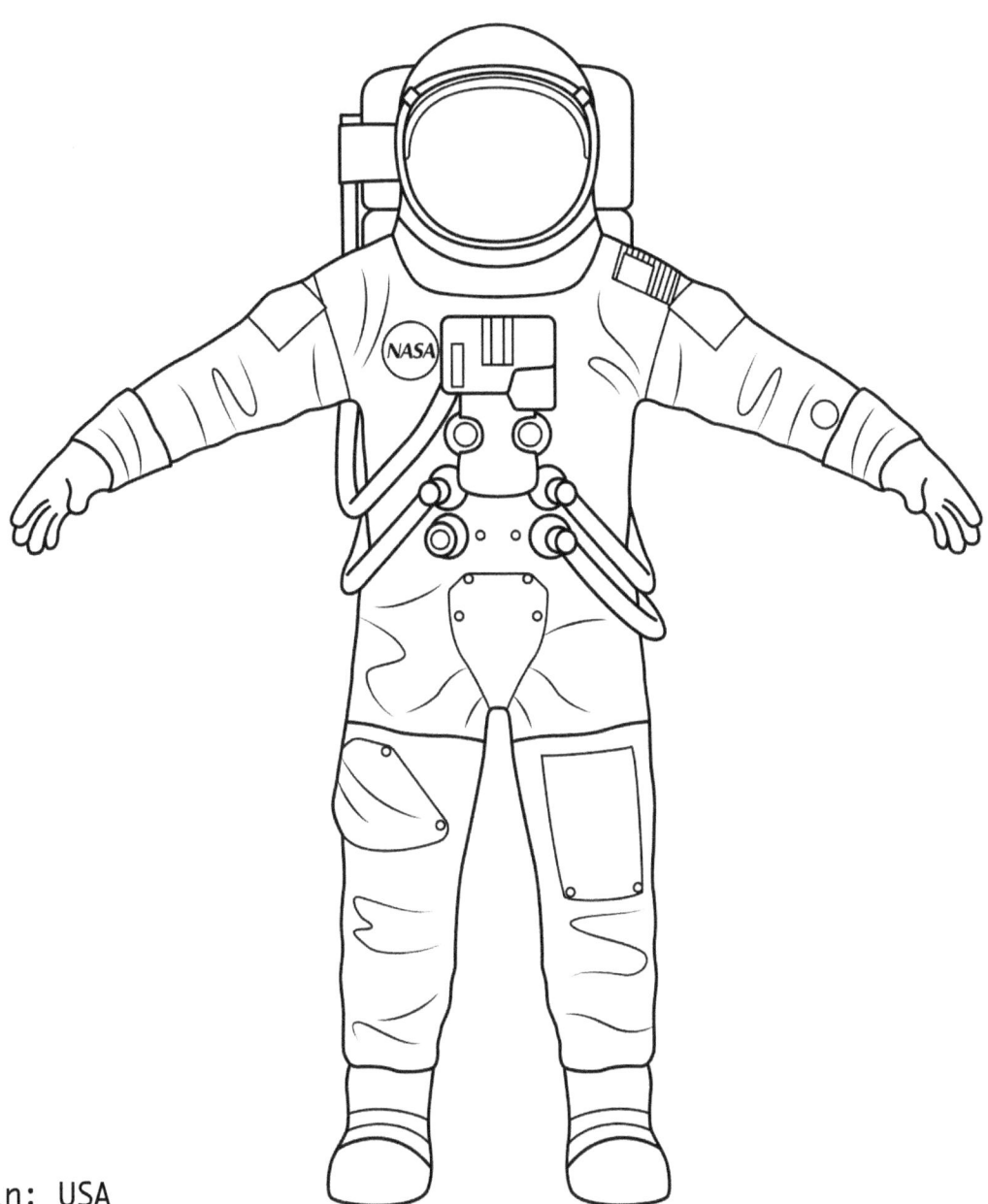

Origin: USA

Date: 1961 - 1972

Missions: Apollo 7-14

Function: Lunar Extra-vehicular activity (EVA)

Fact: Worn during the Apollo 11 lunar landing by Astronauts Neil Armstrong and Buzz Aldrin

Apollo 11 A7L EMU Spacesuit

Apollo 11 A7L EMU Spacesuit

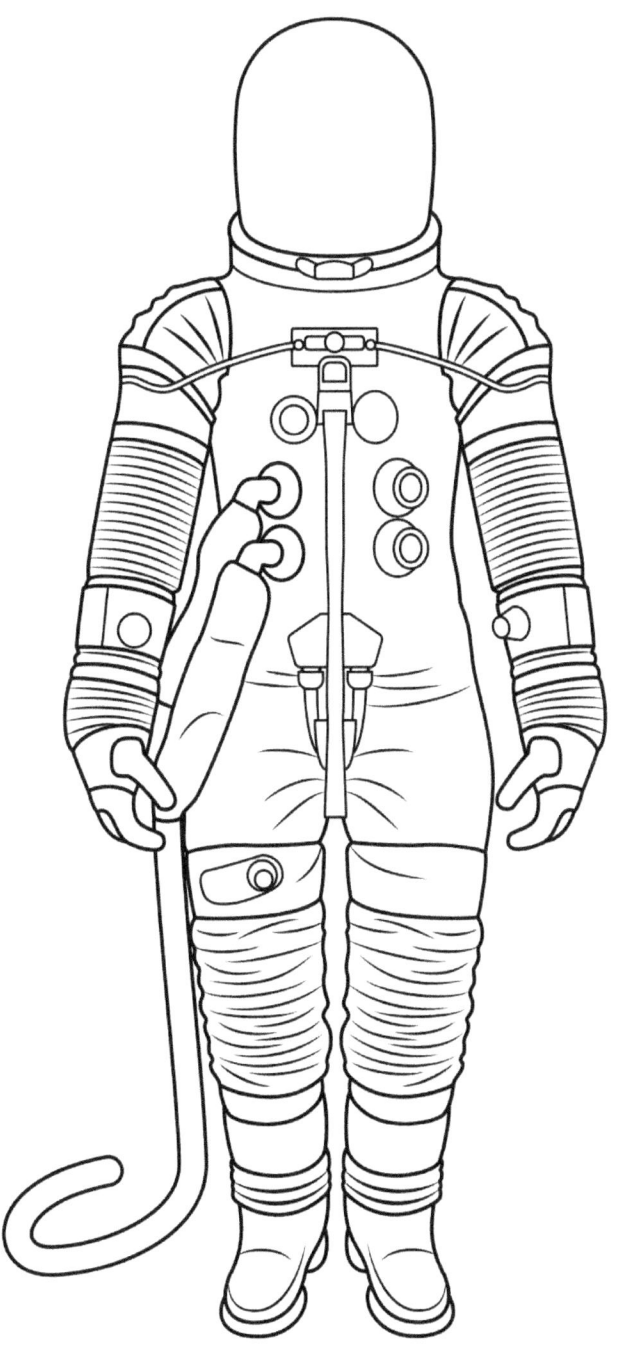

Apollo 11 A7L EMU Spacesuit with the outer layer and visor assembly removed

Apollo 11 A7L EMU Spacesuit

Sokol Spacesuit

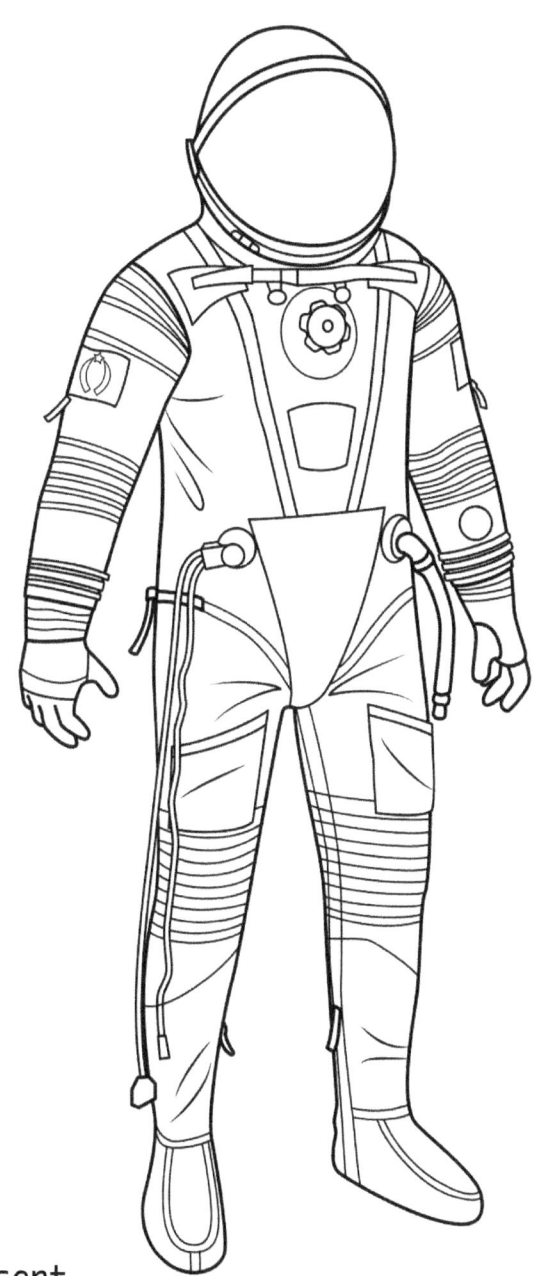

Origin: Russia

Date: 1973 - Present

Missions: Soyuz 12 - Present

Function: Intra-vehicular activity (IVA)

Fact: Worn by crew of the Soyuz Spacecraft during launch and reentry

Sokol Spacesuit

Orlan Spacesuit

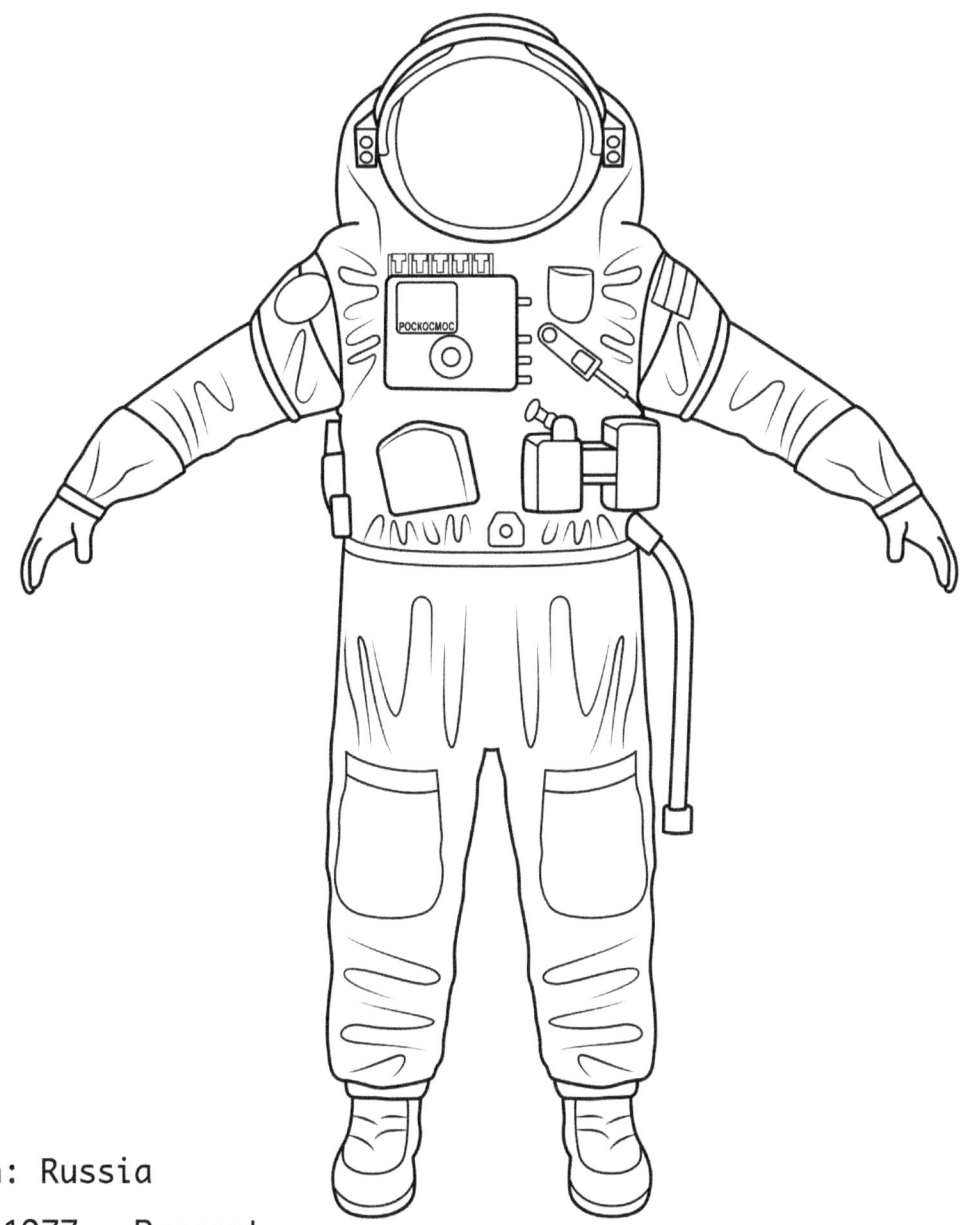

Origin: Russia

Date: 1977 - Present

Missions: Soyuz 26 - Present

Function: Extra-vehicular activity (EVA)

Fact: Seven models of the Orlan suit have been created, the latest model Orlan-MKS is being used on the International Space Station today

Orlan Spacesuit

Shuttle Ejection Escape Spacesuit

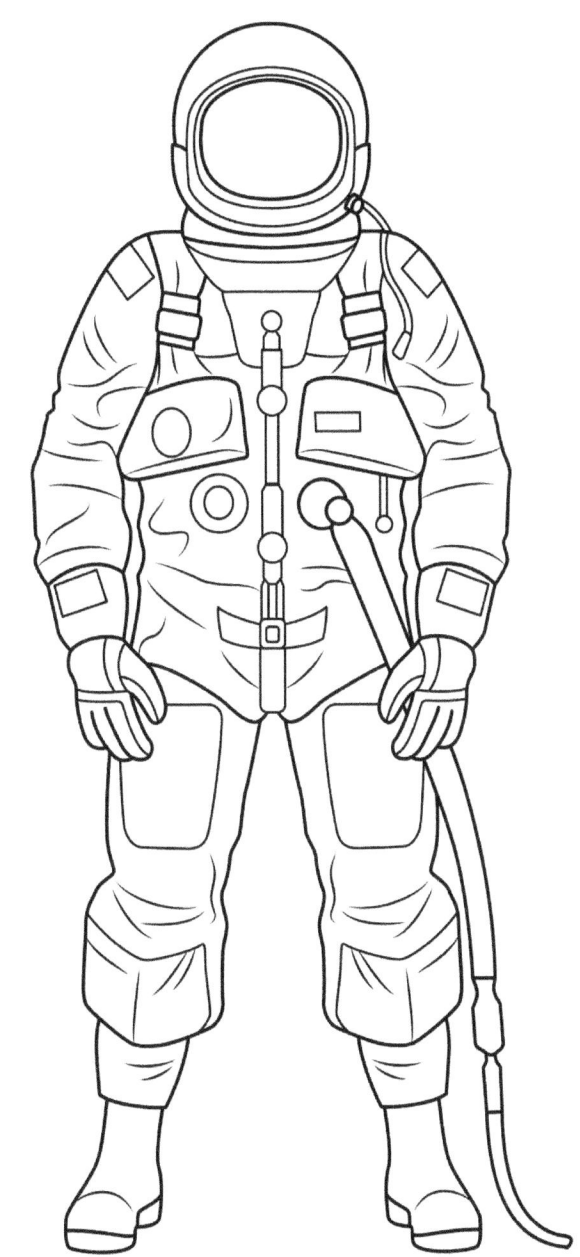

Origin: USA

Date: 1981 - 1984

Missions: STS-1 – STS-4

Function: Intra-vehicular activity (IVA) and Ejection

Fact: Modified version of a U.S. Air Force high-altitude pressure suit

Shuttle Ejection
Escape Spacesuit

Extravehicular Mobility Unit (EMU) Spacesuit

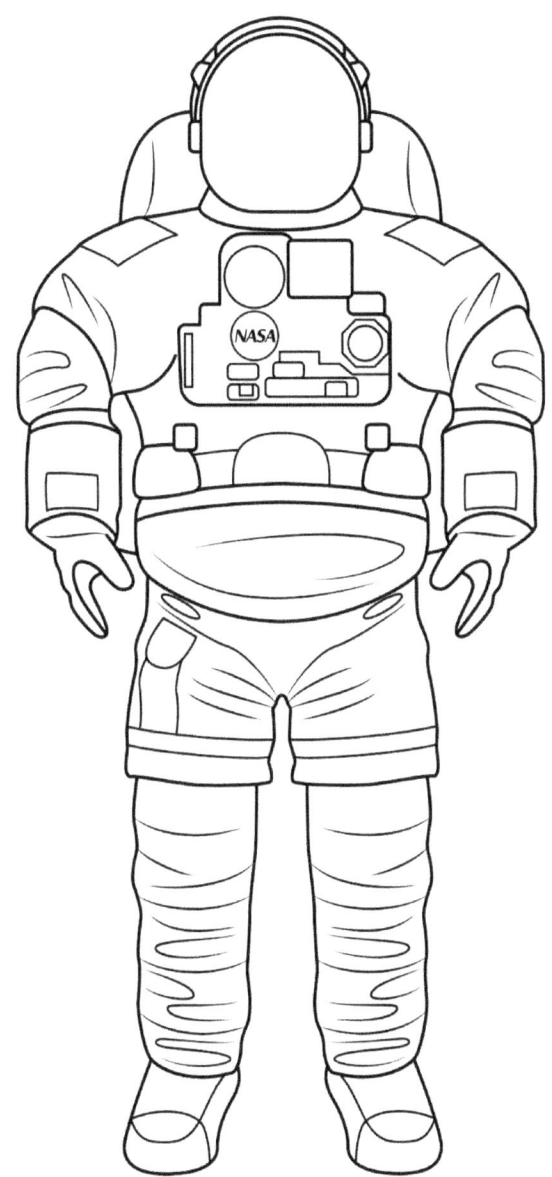

Origin: USA

Date: 1981 - Present

Missions: STS-6 - Present

Function: Extra-vehicular activity (EVA)

Fact: Provides environmental protection, mobility, life support, and communications for astronauts when outside of the spacecraft

Extravehicular Mobility Unit (EMU) Spacesuit

Advanced Crew Escape Spacesuit (ACES)

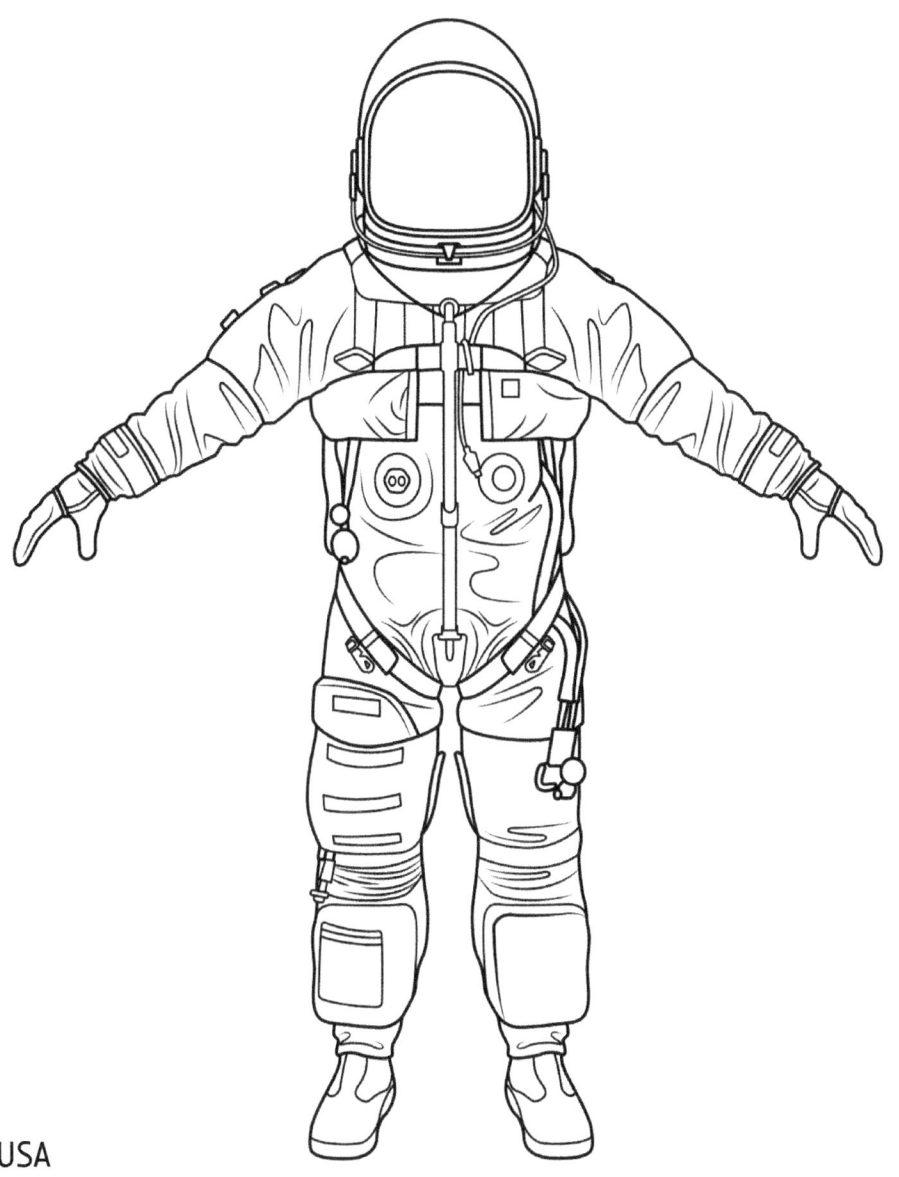

Origin: USA

Date: 1994 - 2011

Missions: STS-64 — STS-135

Function: Intra-vehicular activity (IVA)

Fact: Known as the "Pumpkin Suit" due to its bright orange color allowing astronauts to be spotted if they land in the ocean

Advanced Crew Escape Spacesuit (ACES)

Feitian Spacesuit

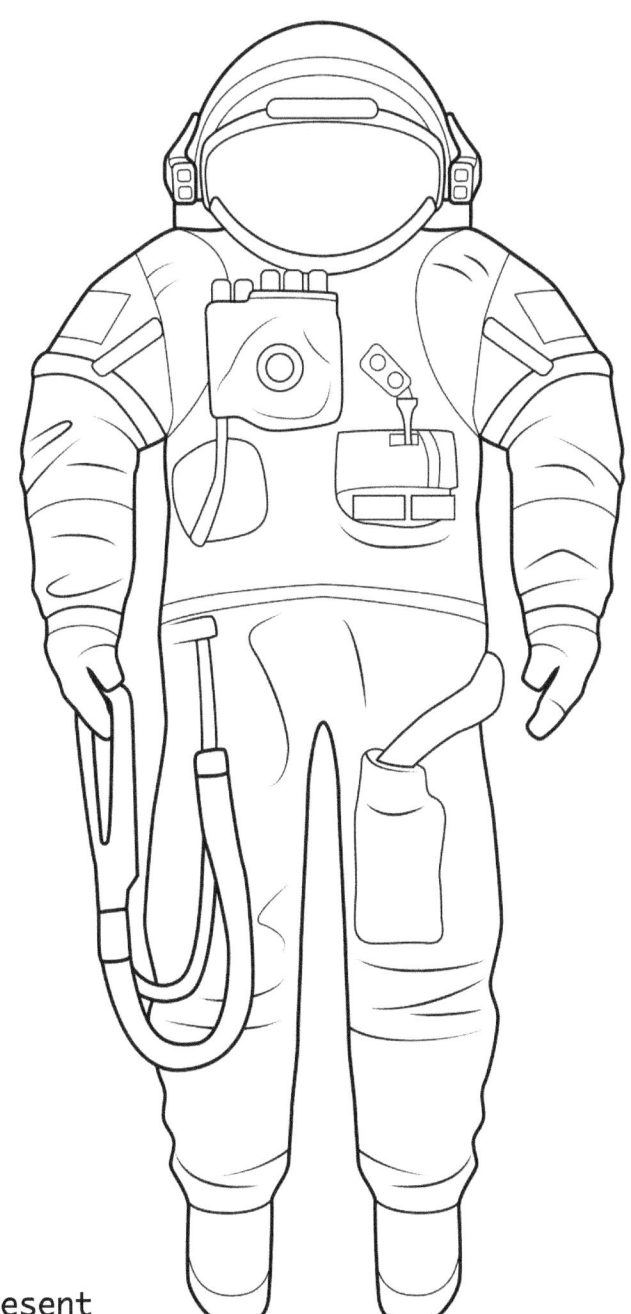

Origin: China

Date: 2008 - Present

Missions: Shenzhou 7

Function: Extra-vehicular activity (EVA)

Fact: Modelled after the Russian Orlan Spacesuit and worn in 2008 for China's first ever spacewalk

Feitian Spacesuit

SpaceX Spacesuit

Origin: USA

Date: 2020 - Present

Missions: Crew Dragon Demo-2 - Present

Function: Intra-vehicular activity (IVA)

Fact: Suit worn by astronauts in the SpaceX Commercial Crew Program

SpaceX Spacesuit

Boeing Starliner Spacesuit

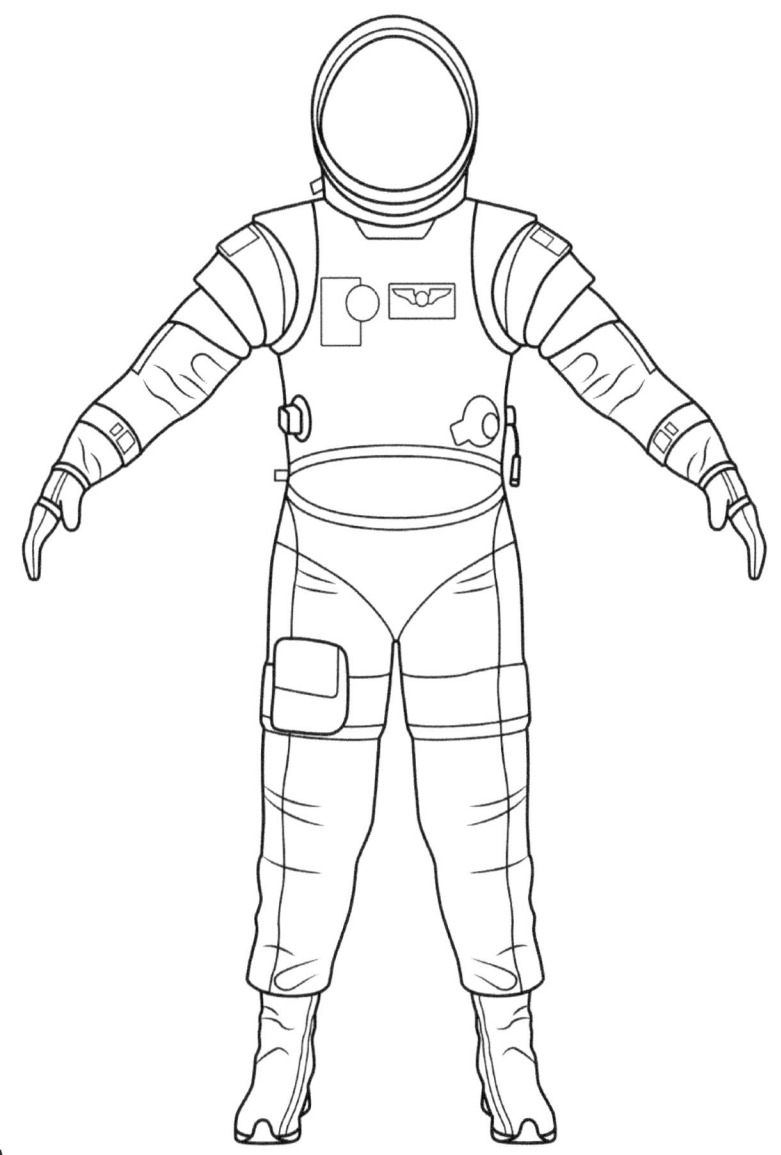

Origin: USA

Date: Proposed 2021

Missions: CST-100 Starliner Commercial Crew

Function: Intra-vehicular activity (IVA)

Fact: Features a soft zip closure helmet contributing to the weight of the suit being approximately 40% lighter than previous IVA suits

Boeing Starliner Spacesuit

Orion Crew Survival System Spacesuit

Origin: USA

Date: Proposed 2024

Missions: Artemis Lunar Missions

Function: Intra-vehicular activity (IVA)

Fact: Designed to enable survival for up to six days, due to the ability to remain pressurized for approximately a week

Orion Crew Survival System Spacesuit

xEMU Spacesuit

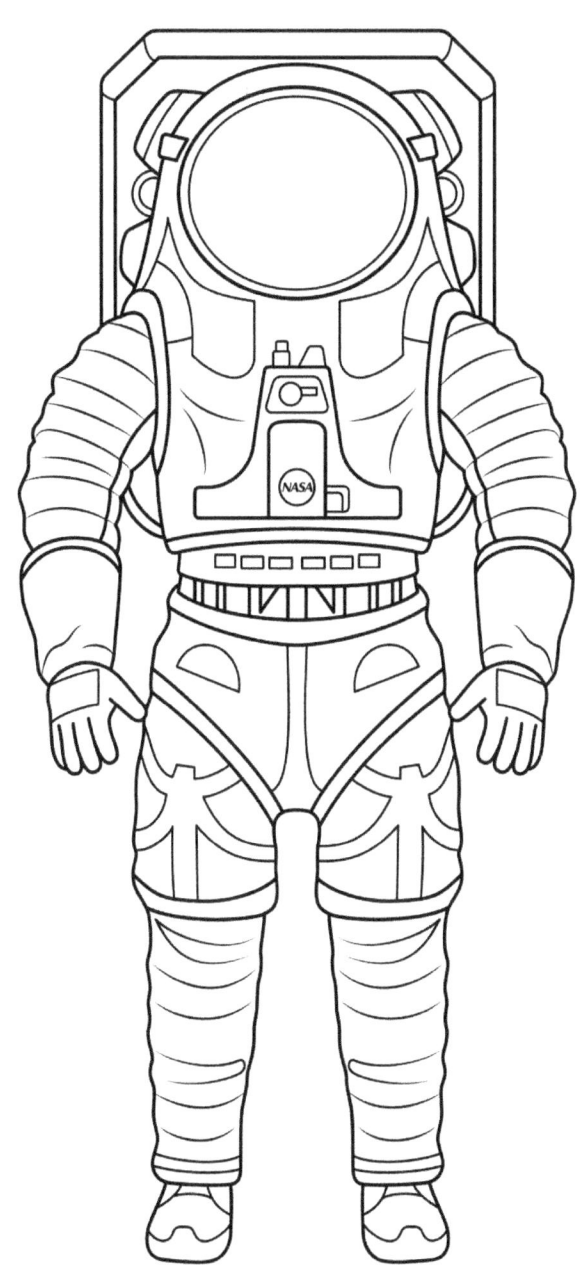

Origin: USA

Date: Proposed 2024

Missions: Artemis Lunar Missions

Function: Lunar Extra-vehicular activity (EVA)

Fact: Proposed suit for the first women to wear on the lunar surface during the Artemis Lunar Missions

xEMU Spacesuit

My Spacesuit Designs

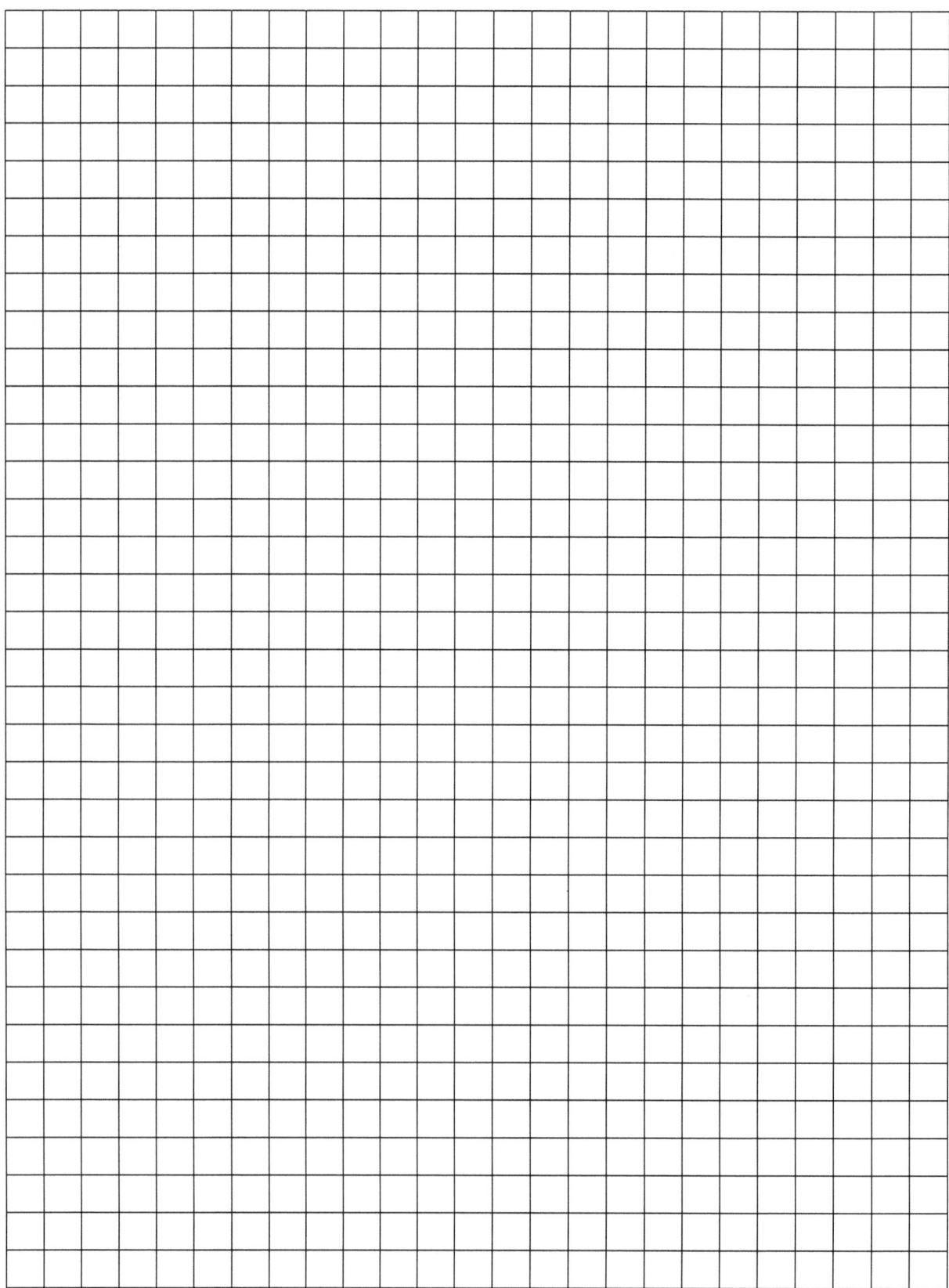

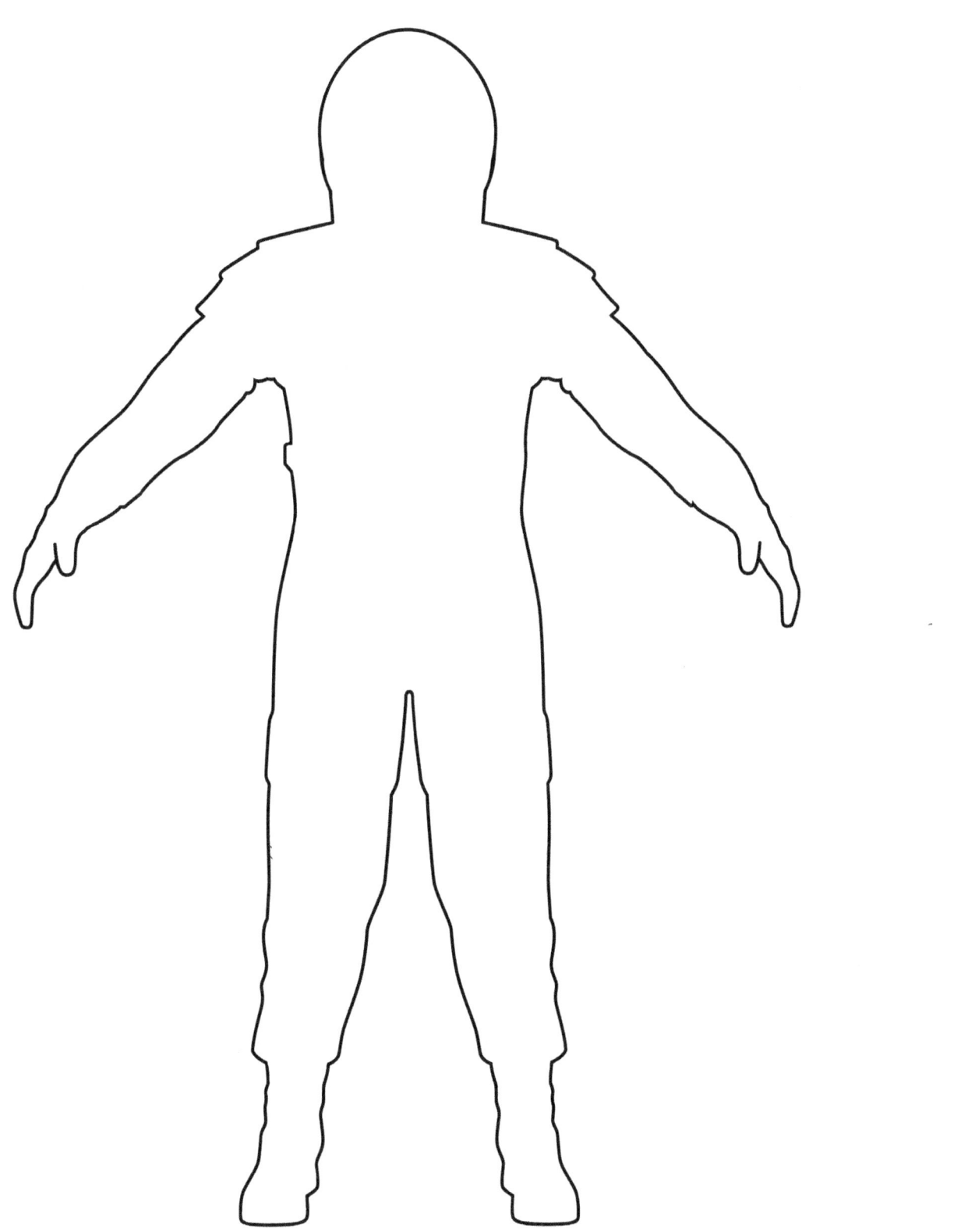

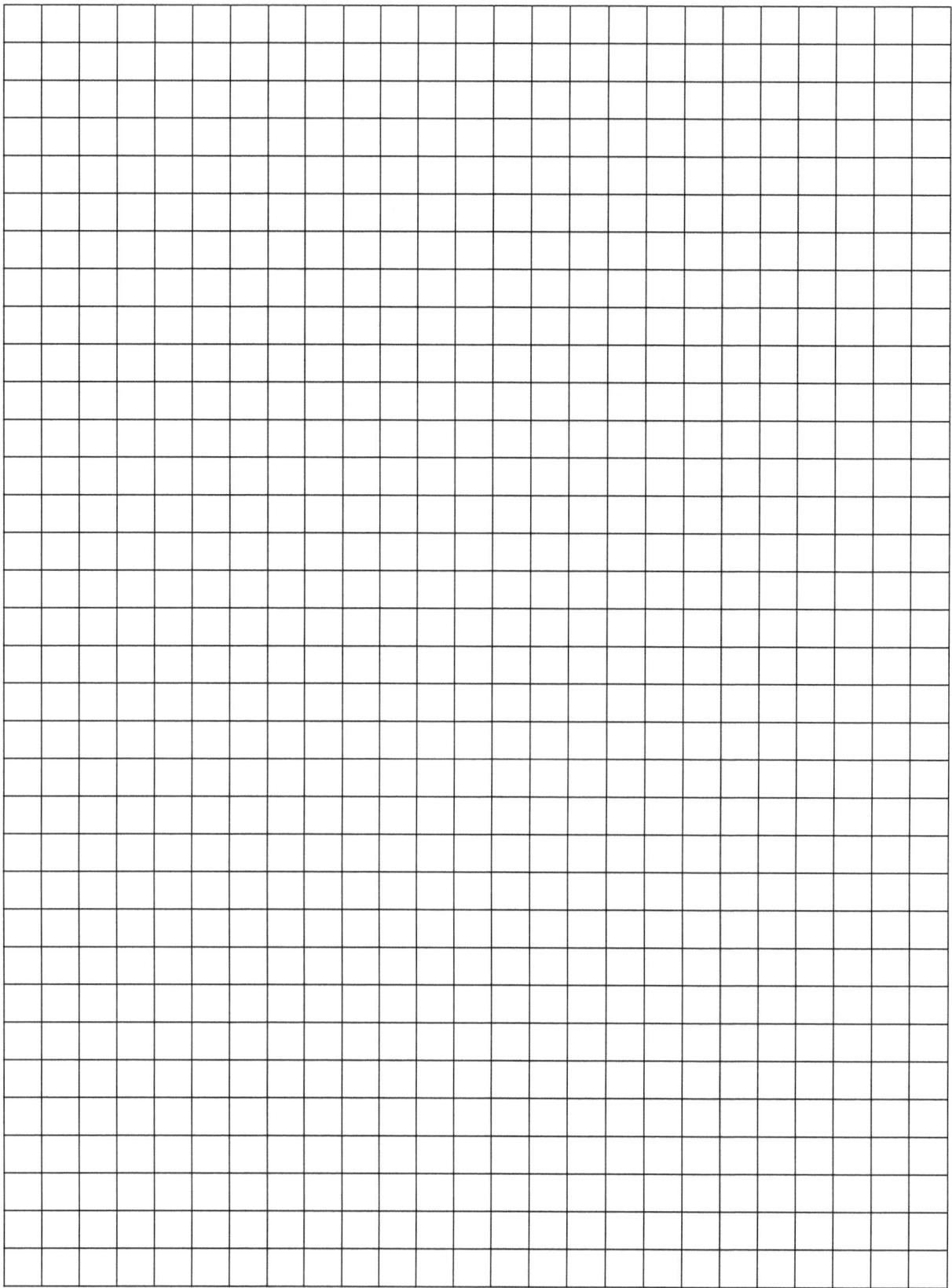

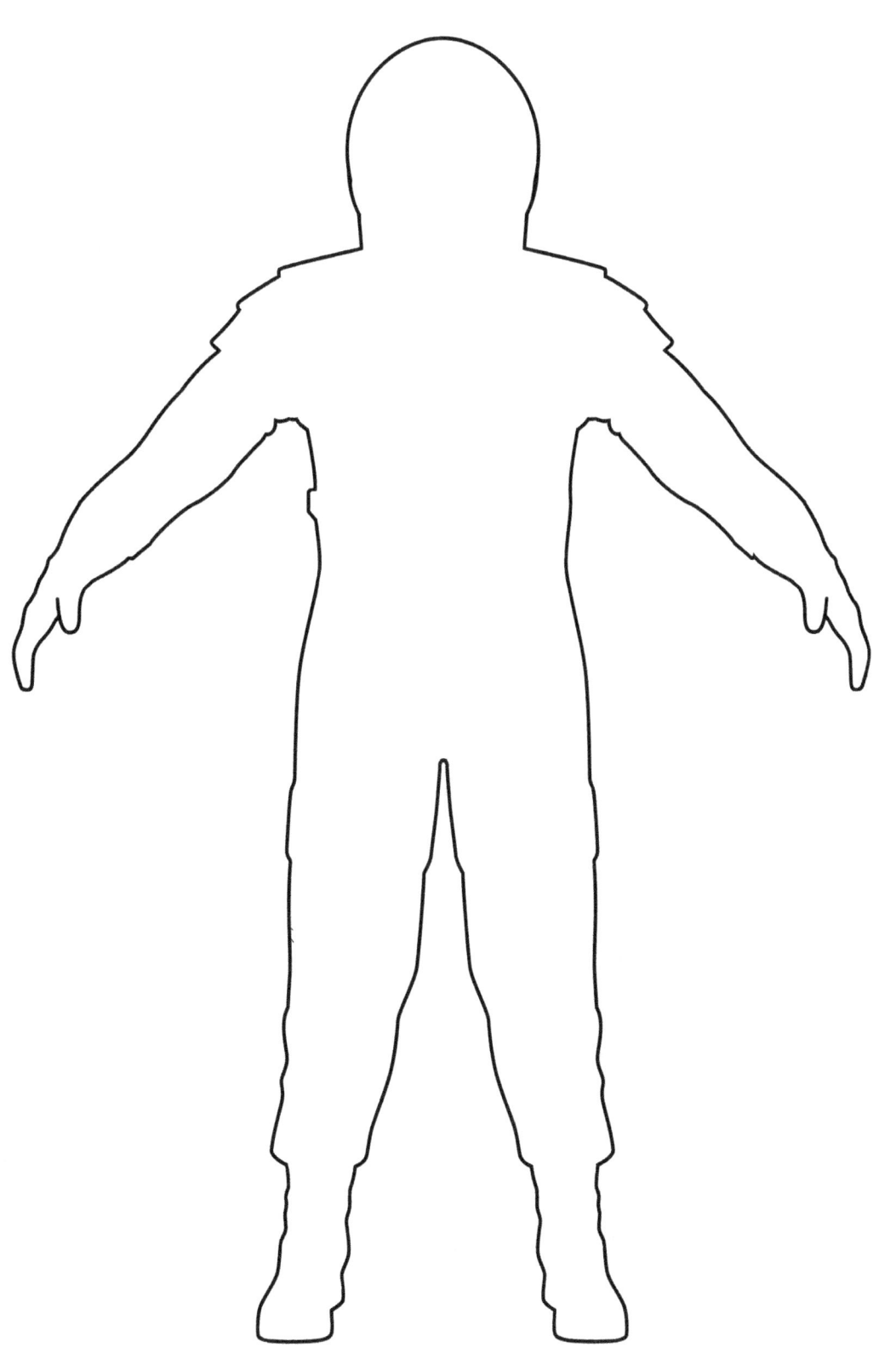

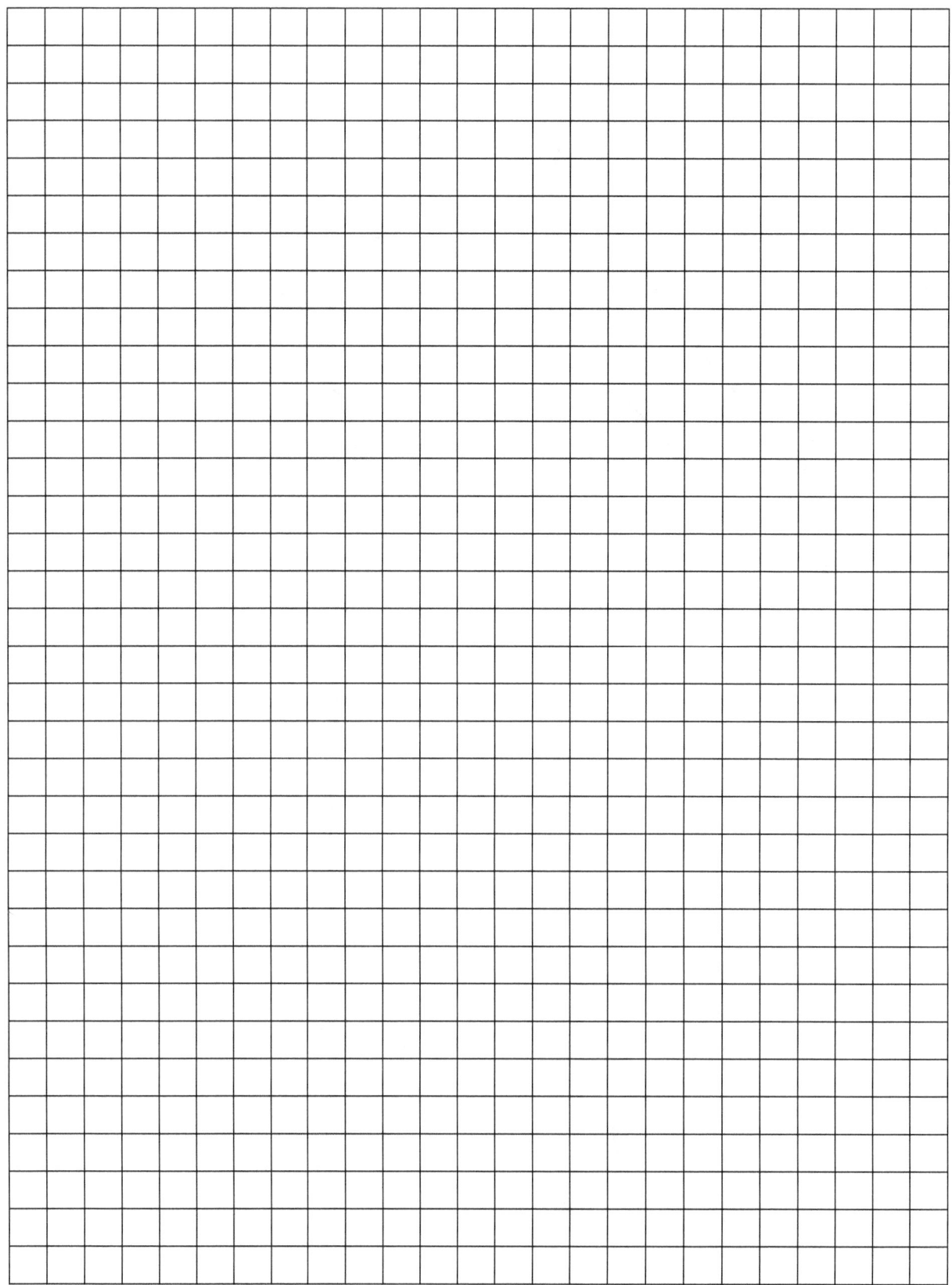

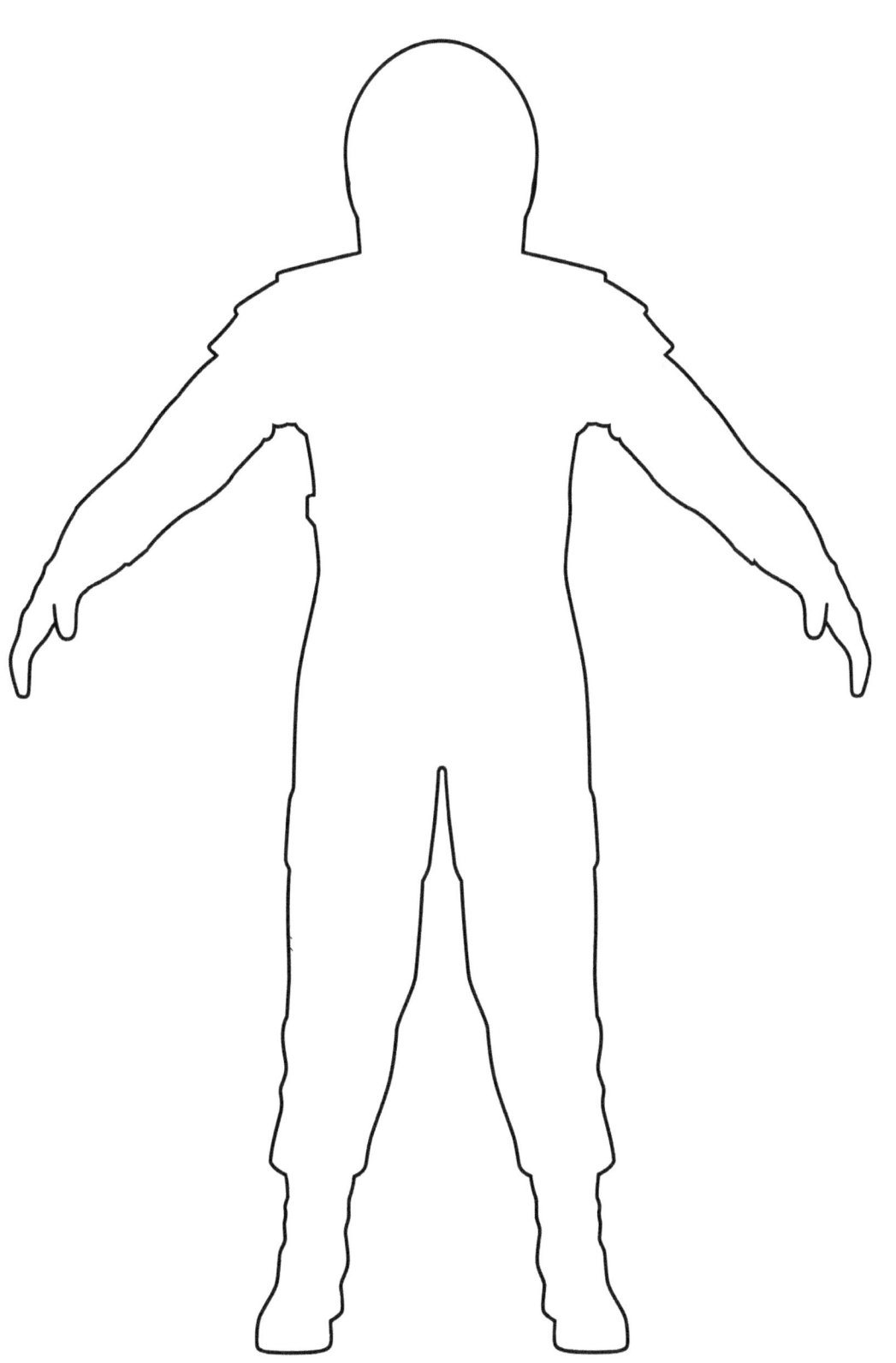

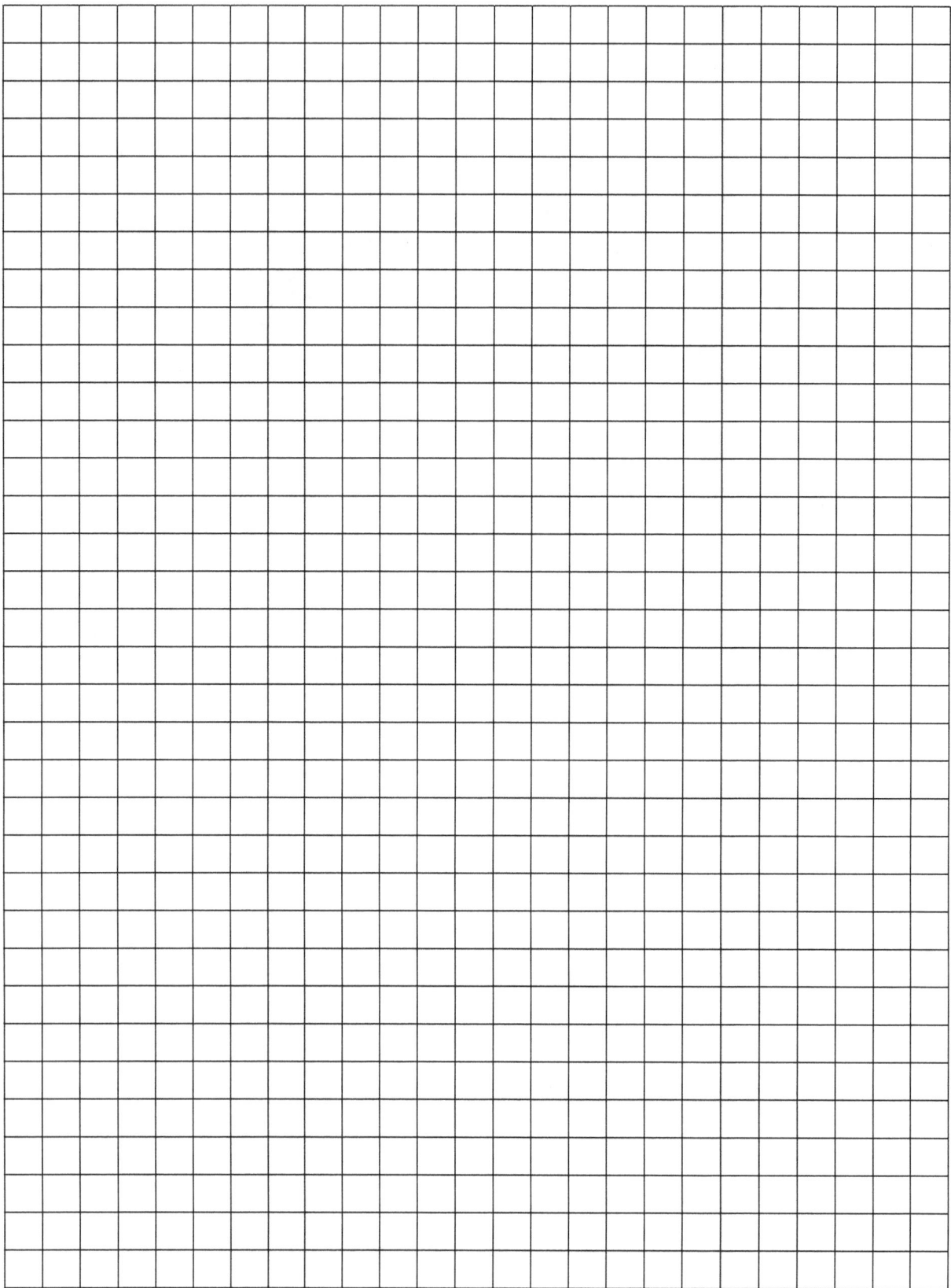

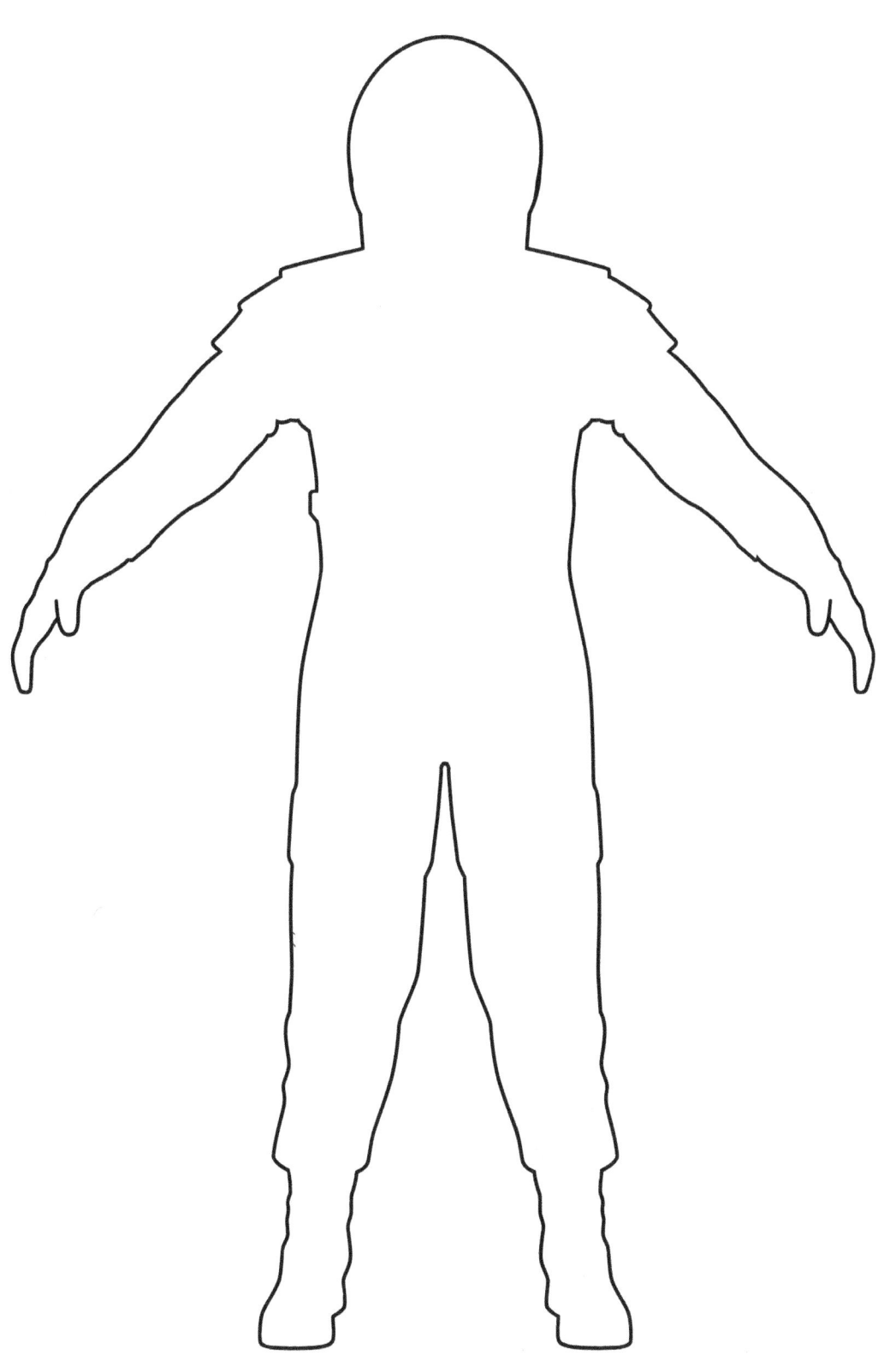

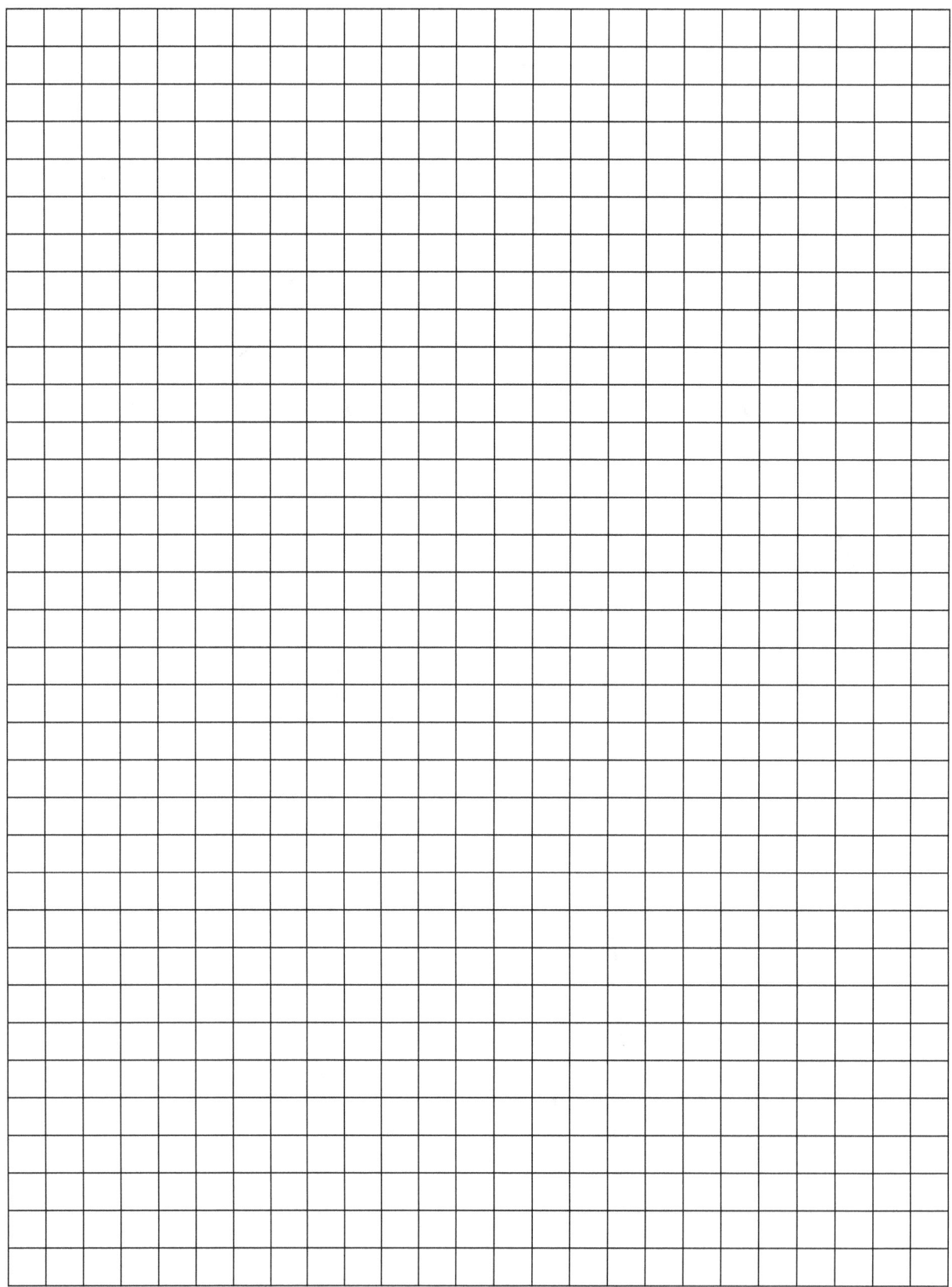

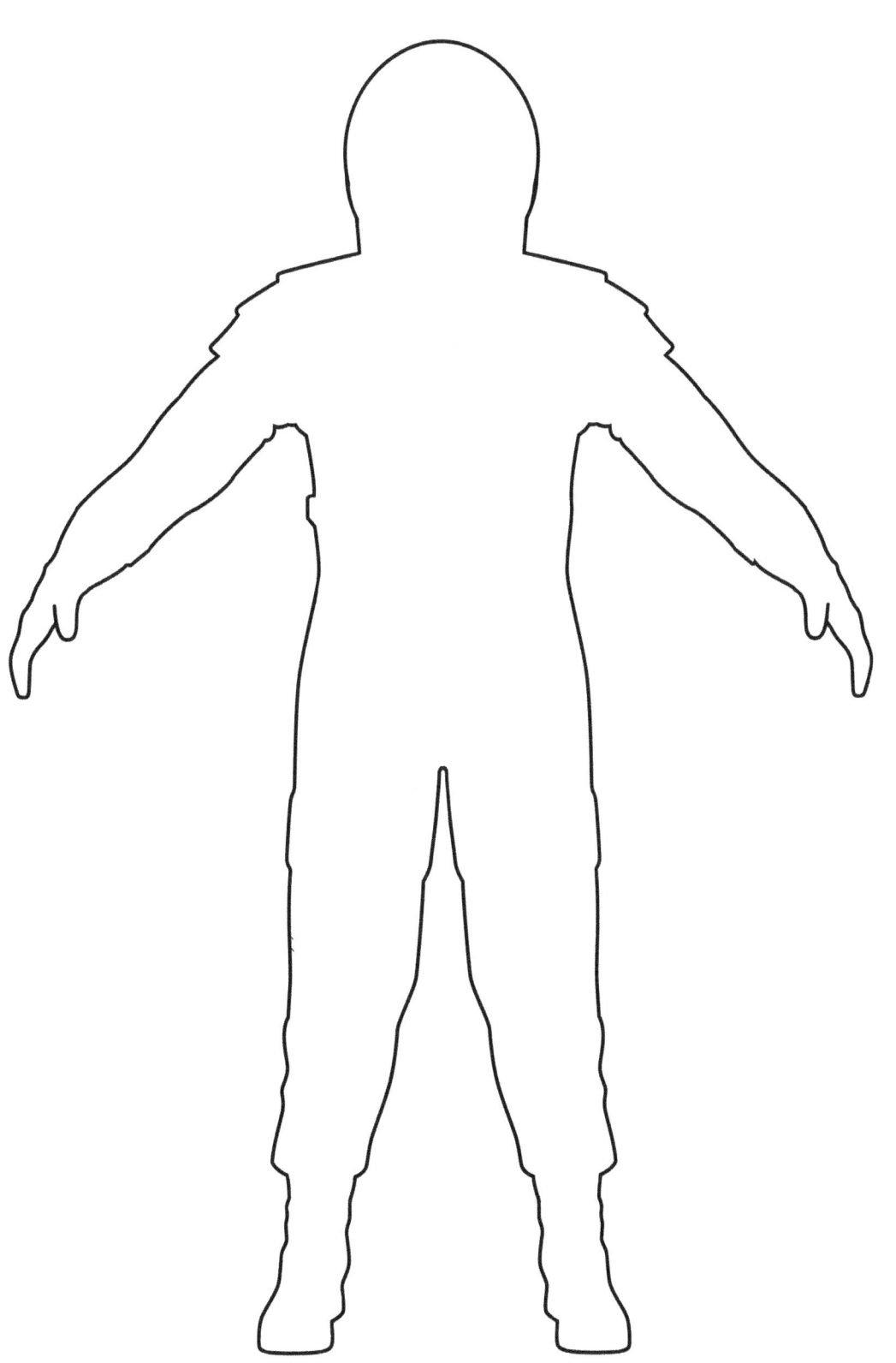

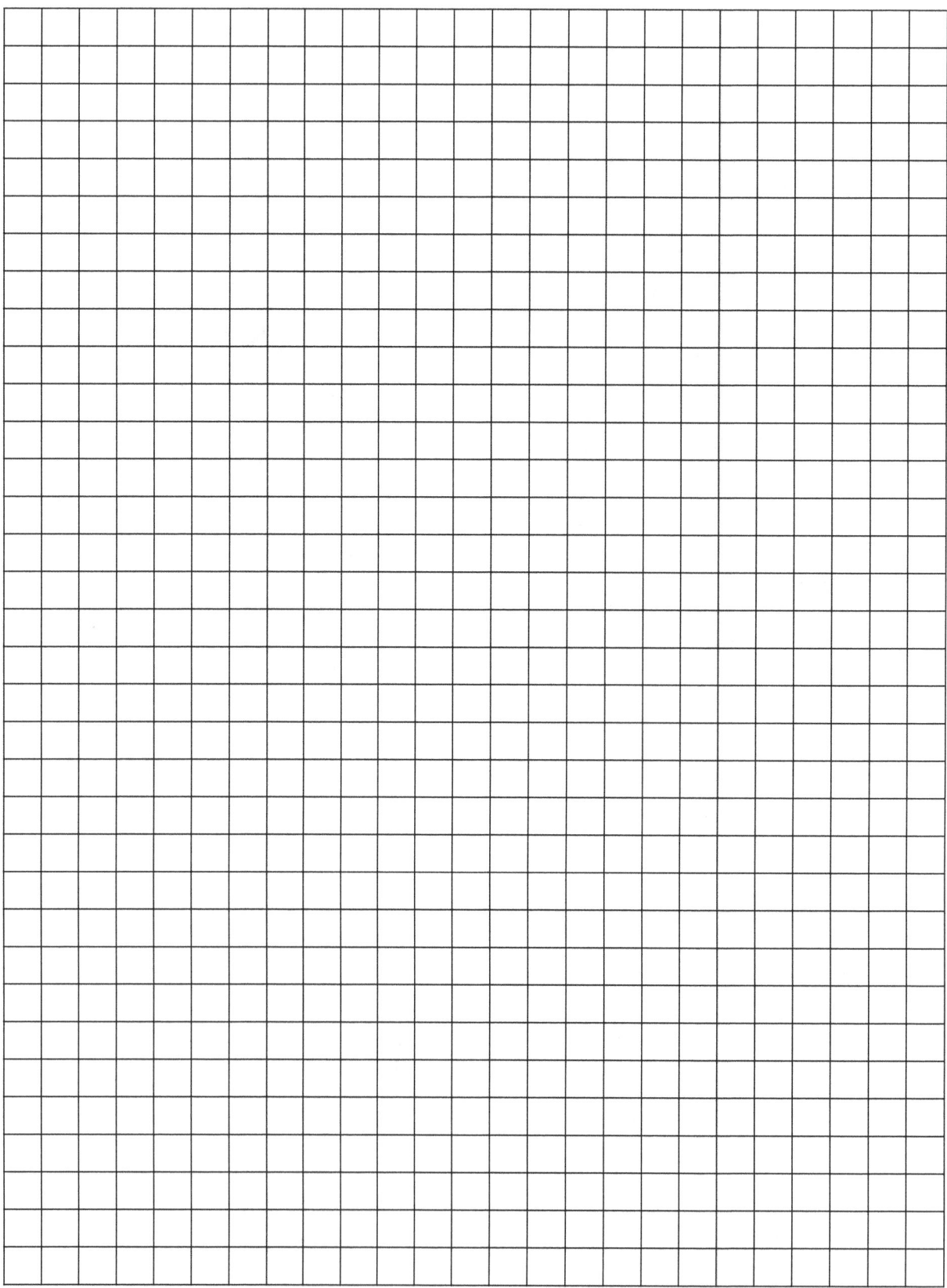

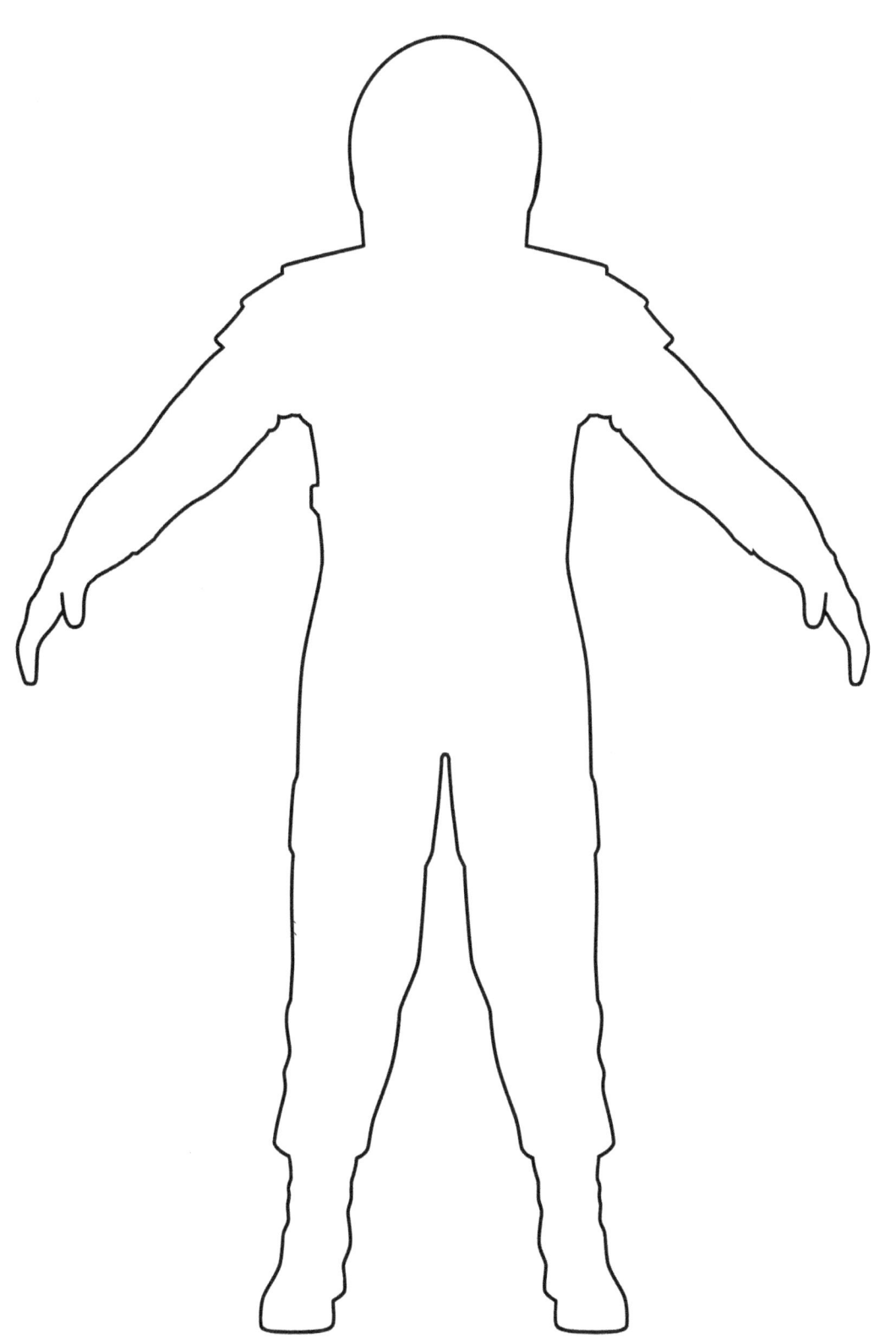

References

1. *Thomas, Kenneth S.; McMann, Harold J. (November 23, 2011). U.S. Spacesuits. Springer Science & Business Media.*

2. *Isaac Abramov & Ingemar Skoog (2003). Russian Spacesuits. Chichester, UK: Praxis Publishing Ltd. ISBN 1-85233-732-X.*

3. *Chen, Lou (September 27, 2008). "Taikonaut Zhai's small step historical leap for China". Xinhua. Archived from the original on October 1, 2008. Retrieved October 1, 2008.*

4. *Boeing.com*

5. *SpaceX.com*

6. *NASA.gov*

www.ingramcontent.com/pod-product-compliance
Lightning Source LLC
Chambersburg PA
CBHW051150290426
44108CB00019B/2678